W9-BGN-484

LIFEVIEWS

Published by Creative Education
123 South Broad Street, Mankato, Minnesota 56001
Creative Education is an imprint of The Creative Company

Art direction by Rita Marshall; Production design by The Design Lab/Kathy Petelinsek
Photographs by Tom Stack & Associates

Library of Congress Cataloging-in-Publication Data
Halfmann, Janet.
Life in a tide pool / by Janet Halfmann; p. cm. — (LifeViews) Includes index
Summary: Briefly describes some of the creatures that live in a tide pool, including periwinkles,
limpets, and sea stars. Suggests several related activities.
ISBN 1-58341-076-7
1. Tide pool animals—Juvenile literature. 2. Tide pool ecology—Juvenile literature. [1. Tide pool animals.
2. Tide pool ecology. 3. Ecology} I. Title. II. Series: LifeViews (Mankato, Minn.)
QL122.2.H345   2000
591.769'9—dc21   99-23823

First Edition

2   4   6   8   9   7   5   3   1

LIFE IN A
TIDE POOL

JANET HALFMANN

PHOTOGRAPHS BY TOM STACK & ASSOCIATES

TWICE EACH DAY ocean tides wash up onto the world's seashores and then flow out again. On many seashores, the outgoing tide leaves behind pockets of water trapped in holes and hollows. These areas, called **tide pools**, can be found on rocky coasts around the world, including North America. Tide pools dance with life. Some of the residents live there all the time, while others arrive with the high tide waters and leave when the tide goes out. Visiting a tide pool at low tide is like peering into a mini-ocean. Life there is abundant, always moving, always changing. And the variety of shapes, colors, and textures is magnificent.

Ocean creatures form an intricate web of life.

Life for creatures in the tide pool is not easy. Like their ocean relatives, tide pool dwellers need the **salty** water of the sea to survive. But the amount of salt in a tide pool is constantly changing, forcing its inhabitants to adapt quickly. The pool gets too salty as the sun dries up the water, and not salty enough as the rain fills it.

The temperature in the tide pool also varies greatly, unlike in the huge ocean where it stays pretty much the same. To escape the sun's heat, **periwinkles** and other tide pool creatures nestle under seaweeds where it is cool. Sea anemones, hydroids, bryozoans, and other tide pool dwellers also must depend on the incoming tide for a fresh supply of food and oxygen.

In this place where the sea meets the land, periwinkle and **limpet** sea snails graze over rocks. Like tiny lawn mowers, they scrape bits of algae off the rocks with an amazing coiled tongue called a **radula**. The radula is long and has hundreds of rows of jagged teeth, which work like a rough file. The periwinkle's tongue is longer than its body!

A great variety of life exists in a tide pool. Sea urchins and snails live side by side. Colorful fish dart in and out of anemone tentacles. And many kinds of seaweed are abundant.

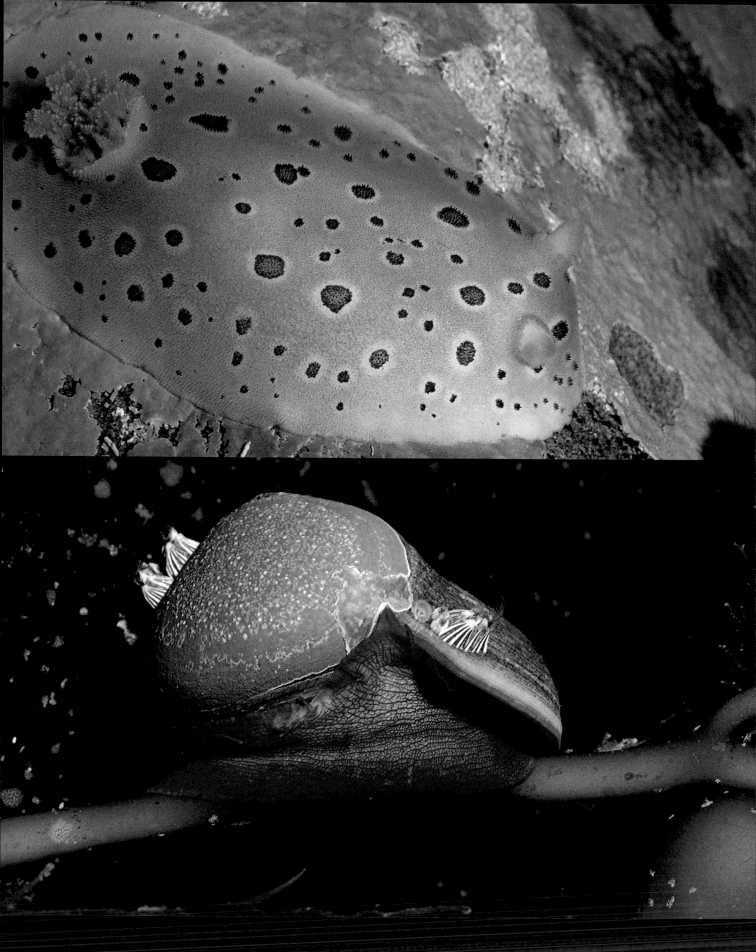

Sea snails are mollusks, soft-bodied animals that usually live in hard shells. A special organ called a **mantle** makes the shell, and the snail adds to it as it grows. The snail moves around on a long, flat foot and leaves slimy trails wherever it goes. It has two eyes and two movable feelers called **tentacles** on its head.

Periwinkles are the size of a fingertip and wear coiled shells. The grayish common periwinkle is one of the most abundant of the 100 species, or kinds. Sometimes the periwinkle crawls on rocks above the tide pool and is uncovered when the tide goes out. Then it crawls inside its shell and glues itself to the rock with **slime**. It snaps shut the lid attached to its foot. Now, it has its own little sea trapped inside its shell to keep it moist until the tide returns.

Common periwinkles mate in the spring and lay their eggs in tiny floating capsules at high tide. The eggs hatch into **larvae**. They are microscopic, meaning they are too tiny to be seen with the naked eye. They swim in the ocean current

Tide pools are filled with many slow-moving creatures. These include the nudibranch, or sea slug (opposite top), and the smooth turbine snail (opposite bottom). The tide brings food within reach of many tide pool dwellers.

with millions of other tiny floating plants and animals called plankton. Larvae don't look anything like their parents. Only after two to three months do they settle on the shore and develop into pinhead-sized periwinkles with shells.

Periwinkles are in danger of being eaten by **gulls** and other birds when the tide is out, and by fish when the tide is in. But their closed-up shells often protect them from enemies that eat them whole. They just pass through unharmed!

The limpet makes a special place on the rocks. It rubs its shell on the rock until it makes a hollow. Then it fits itself into the space. At night and at high tide, it wanders a short distance to graze on **algae**, but in the morning it returns to its home.

The limpet's drab-colored shell looks like a flattened cone or little tent about the size of a quarter or half dollar. Some common tide pool dwellers of the more than 500 species are the mask limpet, the fingered limpet, and the Atlantic plate limpet, also known as the **tortoiseshell** limpet.

Many animals in the world have names that describe them. The keyhole limpet (above) is named for the hole in its shell. The rough limpet (opposite) is exactly that, with a rough-surfaced shell.

Like the periwinkle, the limpet clamps itself against a rock to stay moist and to keep from being washed away by the waves. But it has a mightier grip because of its sloping shell and its groove home. Not even the strongest wave can budge it, and neither can most predators—except the **oystercatcher**. This shorebird whacks the limpet with its long bill to shift it slightly, then inserts its bill and rips the snail's strong foot to break its hold.

Another grazer crawls over rocks at the bottom of the pool. It is the **sea urchin**, a relative of the sea star, with spines sticking out all over its round shell. The 950 species of sea urchins and 1,800 kinds of sea stars, or starfish, belong to a group of spiny-skinned ocean animals called **Echinoderms**. Sea urchins range from less than an inch (2.5 cm) to more than three feet (1 m), but those found in tide pools are about the size of a tennis ball.

The spines of most sea urchins are needle-sharp and sometimes poisonous to keep enemies away. Some of the longest **spines** belong to the long-spined urchin. Sea urchins also use

The American oystercatcher searches for food on the shore.

their spines to wedge themselves tightly into cracks to keep from being washed away. Some colorful kinds found in tide pools are the purple sea urchin and the green sea urchin.

Between its rows of movable spines, the sea urchin has hundreds of water-filled **tube feet**. It uses both its spines and feet for walking and snaring food. The slimy tube feet grip the rocks with suction cups at the ends.

The sea urchin's mouth is on its underside. Its five sharp, pointed teeth move every which way to scrape algae, **sponges**, and other small animals off the rocks. The teeth keep growing, so they never get completely worn down.

Many animals avoid sea urchins, but some have found ways to get around their **bristles**. Sea gulls carry sea urchins into the air and drop them on rocks to break their shells. Then they devour the soft body.

Unlike their grazing sea urchin cousins, most **sea stars** are meat-eaters. The sea star is the major enemy of oysters, clams, and other mollusks. It crawls on top of the mollusk and pulls open the shell with

Sea urchins are a favorite seafood in many Asian cultures. Their hard shells are also dried and sold as decorative items. The brilliant colors of the shell do not fade when the animal dies.

its tube feet. Then the sea star does an amazing thing: it turns its stomach inside out, flipping it out of its mouth and into the mollusk! The **mollusk** is turned into liquid and devoured. Then the sea star pulls its stomach back into its body.

Sea stars are star-shaped, and most have five arms, called rays. Some species have more—up to 50. Hundreds of tube feet with suction cups line the arms. A tiny red **eyespot** at the tip of each arm tells light from dark.

If captured by a predator, the sea star may cast off an arm to escape. Then it grows a new arm. The common **comet** star can regrow an entire body from one arm!

Sea stars come in many colors. Along the Atlantic coast, the northern sea star can be pink, orange, cream, gray, green, blue, or purple. On the Pacific coast, the abundant **ochre star** might be yellow, purple, brown, or orange. The spiny sun star is scarlet with bands of white, pink, yellow, or red.

One of the smaller animals in the tide pool is the **hydroid**.

Hard, dried sea stars are often sold in beach souvenir shops. But the tough covering of the sea star is pliable when the animal is alive. This means it can bend and move easily.

The 2,000 species of hydroids usually grow in colonies, or groups, numbering from just a few to many thousands. Some colonies look like small, bushy plants and can be mistaken for seaweed. Other colonies grow on rocks and even animals.

An individual hydroid is called a **polyp** and is about 0.2 inches (5 mm) tall. It looks like a flower on a stem. The "flower" part has a mouth ringed with tentacles. The tentacles capture tiny animal plankton. Often a little cup surrounds the hydroid's body, such as in the zig-zag wine-glass hydroids. The hydroid pulls itself into the cup to protect itself from enemies, such as the **sea spider**. Hydroids are related to sea anemones and jellyfish.

In some hydroid colonies, such as **snail fur**, polyps have special features and jobs, similar to ant colony or beehive members. Polyps with many tentacles gather food, others with large knobs of stinging cells protect the colony, and those with sac-like organs reproduce.

A new **colony** begins when a hydroid larva attaches itself to a rock. It sprouts a bud, which grows into a second hydroid.

Many kinds of crabs live in tide pools. A type of crustacean, crabs have a hard outer skeleton that protects the soft body inside. Many live inside shells discarded by other animals, including the white-speckled hermit crab (opposite).

More buds follow, and the colony grows. Some kinds of hydroids release special buds that look like jellyfish and are called **hydromedusae**. They are bell-shaped with tentacles and are usually smaller than a quarter.

Bryozoans, which are tinier than the head of a pin, also are mistaken for seaweeds. Their name means "moss animals." Like the hydroids, the 5,000 species of bryozoans usually form colonies, sometimes millions strong. Colonies of the orange-brown **spiral-tufted** bryozoan look like little bushes, while the white lacy-crust bryozoan covers rocks like a lace doily.

An individual bryozoan animal is called a **zooid**. It has a box-like protective case that connects it to all the others in the colony. A ring of hollow tentacles called a **lophophore** surrounds its mouth. The animal sticks the lophophore through a hole in its case, and the tentacles spread out in the water like a funnel. Tiny hairs on the tentacles drive tiny plant and animal plankton through the funnel and into the animal's mouth.

An arrow crab (opposite top) shovels food into its mouth using its claws. The lined chiton (above and opposite bottom) scrapes algae from the surface of rocks and plants. Chitons are living fossils, having existed for millions of years.

Like the hydroids, some kinds of bryozoans have colony members with specific jobs. One special zooid looks like a bird's head. It has a pair of jaws that snap shut on small animals wandering over the colony.

Another animal in the tide pool looks so much like a plant that it's named after the anemone flower. But what looks like a big bloom with colorful petals is actually a meat-eating predator called the **sea anemone**. Its "petals" are stinging tentacles. When an animal brushes against them, the sea anemone strikes. Dozens of **poisoned** darts shoot out, paralyzing and killing the victim. The tentacles then fold over, pushing the prey into a gaping mouth. The sting of an anemone doesn't hurt people, but it feels sticky.

The sea anemone is a polyp, like the hydroid. But it usually lives alone rather than in colonies. Its body is a hollow tube, with a **slimy disk** at the bottom and a mouth with tentacles at the top. Sea anemones fill the tide pool with "blossoms" of all sizes in every color of the **rainbow**, from the emerald

Sea anemones are highly diverse. Some may be as small as a few millimeters, while others may be more than a meter wide. Some are tall and thin; others are flat. Some body textures are soft, and others are tough and leathery.

green of the giant green anemone to the orange stripes of the tiny striped anemone.

Sea anemones pull in their tentacles and become jelly blobs when in danger or exposed by low tide. These unusual creatures don't have many enemies because of their tentacles. But sea slugs, called **nudibranchs**, eat them for their stinging cells. They swallow the cells without firing them, using them for their own defense.

Anemones **multiply** in just about every way possible, by dividing, budding, and growing new anemones from bits accidentally torn from a parent's disk. Sea anemones also lay eggs that develop into pear-shaped larvae. Sometimes the larvae swim off on their own, and other times the parent cares for them until they can fend for themselves. The proliferating anemone often has 30 or more young around its bottom.

The tide pool is home and hiding place, nursery and dining room. Almost every rock and crevice is taken. For many plants and animals, the **challenge** of the ebb and flow of the tides is just a part of everyday life.

Young anemones living on the back of a decorator crab (opposite).

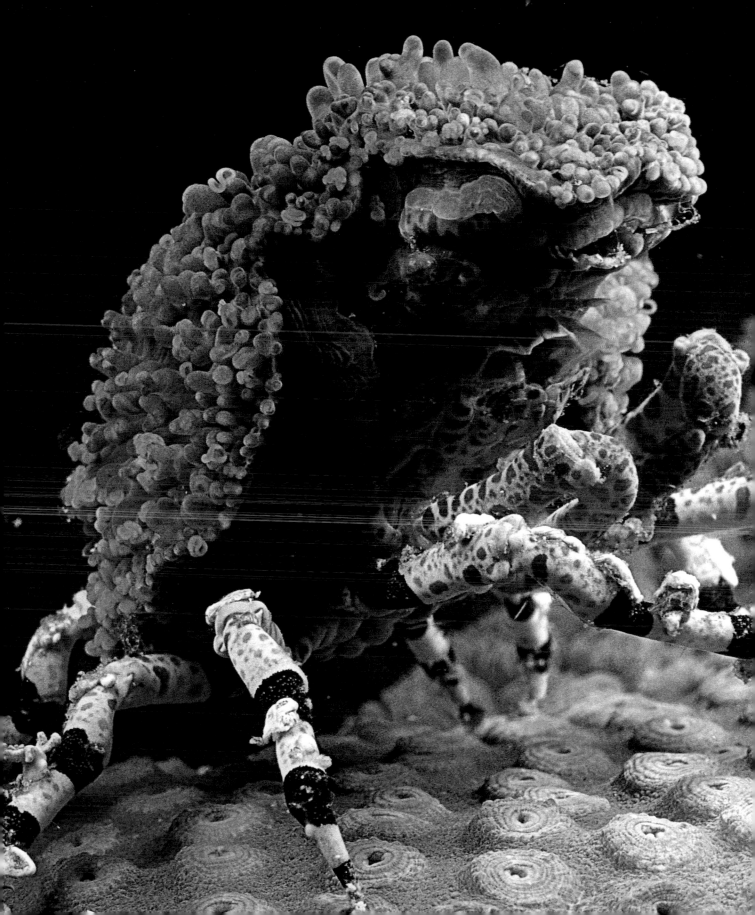

MAKE A TIDE POOL VIEWER

At low tide on rocky seacoasts, you can visit a tide pool and watch the many sea creatures living there. You will be able to see the animals under the water more clearly with a viewer.

Before you go, find out the best time to visit the tide pool. Take an adult with you, and keep a close eye on the water because the tide can move in quickly.

You Will Need
- Half-gallon cardboard milk carton
- Clear heavy plastic wrap
- Waterproof plastic tape or strong rubber band
- Scissors

What To Do
1. Cut off the top and bottom of the milk carton.
2. Cover one end with clear plastic wrap so that it extends about halfway up the carton.
3. Pull the plastic wrap tight and tape around the sides with waterproof tape, or secure it with a strong rubber band.
4. Put the wrapped edge into the tide pool about two inches (5 cm). Look through the other end.

Watch for some tide pool creatures to be staring back at you!

WHAT'S THAT SHELL MADE OF?

Many sea animals, such as periwinkles and barnacles, make shells to protect themselves from enemies and to keep them moist when they're out of water. What are their shells made of? You can easily find out.

You Will Need

- Seashell
- Fingernail file
- Plastic lid
- Vinegar

What To Do

1. Scrape the seashell with the fingernail file.
2. Put the powder from the shell in a plastic lid.
3. Pour vinegar into the lid.

If the mixture fizzes, the shell is made of calcium carbonate, also known as chalk or limestone. The mixture fizzes because of a chemical reaction. When an acid such as vinegar is added to calcium carbonate, a gas called carbon dioxide is released. Carbon dioxide is part of many seashells. When sea animals make their shells, they take carbon dioxide from the water. This is good for Earth's environment because it reduces the amount of carbon dioxide on the planet.

WILL IT FLOAT?

Some animals and plants can live only in salt water, which is found in oceans and seas. Others can survive only in the fresh water of lakes, ponds, and rivers. Tide pool dwellers have to be able to adjust to water with different amounts of salt. When it rains, the tide pool becomes less salty. When the sun dries up some of the water, the tide pool gets more salty. You can do an experiment to see how the amount of salt in water affects floating.

You Will Need

- Pint jar
- Salt
- Water
- Egg
- Tablespoon

What To Do

1. Fill the jar about two-thirds full with tap water.
2. Gently set the egg into the water. Does it float? It shouldn't.
3. Remove the egg and add 3 tablespoons of salt, stirring until it's dissolved.
4. Gently set the egg into the water. Does it float? It should.

Now, you can see why it's easier for animals and people to float in salty sea water than in fresh lake water. The salt mixture in this experiment is saltier than most sea water, but not as salty as the Dead Sea, a huge lake located between Israel and Jordan. The Dead Sea is so salty that no more salt can be dissolved in it, and hardly anything grows there.

LEARN MORE ABOUT TIDE POOLS

Birch Aquarium at Scripps
9500 Gilman Drive
Department 0207
La Jolla, CA 92093-0207
http://aqua.ucsd.edu/SBAM2/
sbam_000.html

Department of Invertebrate Zoology
National Museum of Natural History
Smithsonian Institution
Washington, D.C. 20560
http://www.nmnh.si.edu/departments/
invert.html

Marine Biological Laboratory
7 MBL Street
Woods Hole, MA 02543
http://www.mbl.edu/new.index.html

The Monterey Bay Aquarium
886 Cannery Row
Monterey, CA 93940
http://mbayaq.org/index.htm

Mote Marine Laboratory
1600 Ken Thompson Parkway
Sarasota, FL 34236
http://www.marinelab.sarasota.fl.us/

National Aquarium in Baltimore
Pier 3 at 501 E. Pratt Street
Baltimore, MD 21202
http://www.aqua.org/

Phycological Society of America
Membership Office
P.O. Box 1897
Lawrence, KS 66044-8897
(Note: phycology is the study of algae)
http://jupiter.phy.ohiou.edu/psa/

Vancouver Aquarium Marine
 Science Center
P.O. Box 3232
Vancouver, British Columbia
Canada V6B 3X8
http://www.vanaqua.org/index2.htm

INDEX

Tide pools are home to many colorful living ornaments.